THE BUSINESS SUCCESS GUIDE TO CAMEL HUSBANDRY

Best Practices For Breeding, Nutrition, Health Management, And Profitability In Modern Farming

RICHMOND HAMILL

© 2024 [RICHMOND HAMILL]. All rights reserved.

Except for brief quotations included in critical reviews and certain other noncommercial uses allowed by copyright law, no part of this book may be reproduced, distributed, or transmitted in any form or by any means, including photocopying, recording, or other electronic or mechanical methods, without the publisher's prior written permission.

Disclaimer

The information presented in this book is based on the author's personal knowledge and understanding of livestock management. The author is not affiliated with any association, company, business, or individual in the livestock industry. All content is provided for informational purposes only and should not be considered as professional advice. Readers are encouraged to seek professional guidance and conduct their own research before making any decisions based on the information contained in this book. The author and publisher disclaim any liability for any adverse effects or consequences resulting from the use of the information contained herein.

Table of Contents

CHAPTER ONE 9

Introduction To Camel Husbandry 9
Overview Of Camel Husbandry 9
Importance Of Camels In Various Cultures....10
Basic Camel Anatomy And Physiology 12
Historical Significance Of Camels 13
Benefits Of Raising Camels.............................14

CHAPTER TWO 17

Selecting The Right Camel 17
Choosing Camel Breeds For Specific Needs 17
Key Characteristics To Look For......................19
Health Indicators In Camels 20
Understanding Camel Age And Gender...........21
Tips For Buying Camels 23

CHAPTER THREE..25

Camel Housing And Environment 25

Designing A Camel Pen Or Shelter 25

Necessary Equipment And Supplies 26

Maintaining Cleanliness And Hygiene 28

Importance Of Space And Ventilation 29

Protecting Camels From Extreme Weather 30

CHAPTER FOUR 33

Feeding And Nutrition 33

Essential Nutrients For Camels 33

Types Of Camel Feed And Supplements 35

Feeding Schedule And Quantity 36

Common Nutritional Issues And Solutions 37

Water Requirements And Management 39

CHAPTER FIVE 41

Camel Health And Veterinary Care 41

Common Health Issues In Camels 41

Routine Health Checks And Vaccinations 43

Recognizing Signs Of Illness 44

Emergency First Aid For Camels 46

Finding And Working With A Veterinarian 47

CHAPTER SIX..49

Basics Of Camel Reproduction 49

Breeding Techniques And Practices 50

Gestation And Birth Process..........................52

Care For Pregnant And Newborn Camels........53

Managing Camel Herds For Optimal Reproduction 55

CHAPTER SEVEN...57

Training And Handling Camels 57

Basic Training Techniques For Camels 57

Building Trust And Handling Skills 58

Training For Specific Tasks (E.G., Riding, Packing) 59

Addressing Behavioral Issues 60

Safety Tips For Handling Camels 61

CHAPTER EIGHT 63

Camel Products And Uses 63

Milk Production And Processing 63

Meat And Wool Utilization 65

Leather And Craft Products 66

Economic Value Of Camel Products 68

Marketing And Selling Camel Products 69

CHAPTER NINE 73

Challenges And Solutions In Camel Husbandry
.... 73

Common Challenges Faced By Camel Owners 73

Solutions And Best Practices 75

Adapting To Climate And Environmental Changes 77

Resources For Ongoing Support 78

CHAPTER TEN 81

Legal And Ethical Considerations 81

Understanding Local Regulations And Laws ..81

Ethical Treatment Of Camels 83

Record-Keeping And Documentation 84

Responsible Breeding Practices 85

Engaging With Camel Husbandry Communities 87

Frequently Asked Question And Nswers. 89

CONCLUSION 96

THE END 100

ABOUT THIS BOOK

This book Camel Husbandry offers an invaluable resource for both novice and experienced camel farmers, providing comprehensive insights into every facet of camel care and management. It begins by delving into the essential aspects of camel husbandry, highlighting the significance of camels across various cultures and their historical importance. This book explores camel anatomy and physiology, helping readers understand the benefits of raising these remarkable animals, from their economic contributions to their role in traditional societies.

Selecting the right camel is crucial for successful husbandry, and this book guides readers through the process of choosing camel breeds that align with their specific needs. It provides practical advice on identifying key characteristics, assessing health indicators, and understanding camel age and gender.

This foundation ensures that readers make informed decisions when purchasing camels, setting the stage for effective management.

Creating a suitable living environment is essential for camel welfare. This book offers detailed guidance on designing camel pens and shelters, including necessary equipment and supplies. It emphasizes the importance of cleanliness, space, and ventilation, and provides strategies for protecting camels from extreme weather conditions, ensuring their comfort and health.

Feeding and nutrition are addressed with equal thoroughness. This book outlines essential nutrients for camels, various types of feed and supplements, and optimal feeding schedules. It also covers common nutritional issues, water management, and practical solutions to ensure camels receive a balanced diet that supports their health and productivity.

Camel health and veterinary care are paramount, and this guide covers common health issues, routine checks, and vaccinations. It includes advice on recognizing signs of illness, providing emergency first aid, and working effectively with veterinarians. This section equips readers with the knowledge to maintain the well-being of their camels and address health concerns promptly.

Breeding and reproduction are fundamental aspects of camel husbandry. This book discusses camel reproduction basics, breeding techniques, and the care required during gestation and birth. It also offers strategies for managing camel herds to achieve optimal reproductive success, ensuring sustainable growth and productivity.

Training and handling camels are covered with practical advice on basic training techniques, building trust, and handling skills. This book includes methods for training camels for specific tasks, such as riding or packing, and provides tips for addressing behavioral issues, emphasizing safety in handling practices.

The economic potential of camel products is explored in detail, including milk production, meat, wool, and leather. This book discusses processing methods, the economic value of camel products, and marketing strategies, offering insights into how camel products can be effectively utilized and sold.

Addressing challenges in camel husbandry, This book identifies common problems and offers solutions and best practices. It includes case studies and success stories, providing real-world examples of how to adapt to climate and environmental changes. Resources for ongoing support are also highlighted, ensuring that readers have access to continuous learning and problem-solving tools.

Finally, This book covers legal and ethical considerations, including understanding local regulations, ensuring the ethical treatment of camels, and maintaining proper documentation. It advocates for responsible breeding practices and encourages engagement with camel husbandry communities, fostering a culture of care and compliance in camel farming.

Overall, Camel Husbandry is an essential guide for anyone involved in or interested in camel farming, offering a wealth of knowledge to enhance practices, ensure the well-being of camels, and maximize the benefits of camel husbandry.

CHAPTER ONE

Introduction To Camel Husbandry

Overview Of Camel Husbandry

Camel husbandry, the practice of raising and managing camels, is a vital agricultural activity in many regions, particularly in arid and semi-arid areas. Camels are highly valued for their adaptability to harsh climates, their ability to travel long distances, and their multifunctional uses. They are often referred to as "ships of the desert" due to their importance in desert transportation and their capacity to carry heavy loads across challenging terrains.

In camel husbandry, the primary focus is on ensuring the health, productivity, and well-being of the camels.

This involves understanding their dietary needs, housing requirements, and healthcare. Camel husbandry practices include breeding, feeding, milking, and managing their overall health. Proper management techniques not only enhance the productivity of camels but also improve their longevity and quality of life.

For beginners, camel husbandry starts with familiarizing oneself with the specific needs of camels. This includes learning about their unique dietary requirements, their social behavior, and their physical care. Ensuring that camels have access to adequate nutrition, clean water, and suitable living conditions is crucial for successful husbandry.

Importance Of Camels In Various Cultures

Camels hold significant cultural and economic importance in many societies, especially in regions such as the Middle East, North Africa, and parts of

Asia. In these cultures, camels are not just a means of transportation but also symbols of wealth and status. They are integral to traditional lifestyles and have been used for centuries in trade, travel, and as part of cultural rituals.

In nomadic and pastoralist communities, camels are often considered the most valuable asset. They provide essential resources such as milk, meat, and hides, which are crucial for survival in harsh environments. Camel milk, in particular, is highly nutritious and a staple in many diets. Camels are also used in traditional ceremonies and festivals, reflecting their deep cultural significance.

Understanding the cultural context of camel husbandry helps in appreciating the role camels play beyond their practical uses. For those involved in camel husbandry, respecting and incorporating cultural practices and traditions can enhance their relationship with the animals and improve the overall effectiveness of their husbandry practices.

Basic Camel Anatomy And Physiology

To effectively manage camels, it is essential to understand their basic anatomy and physiology. Camels have several distinctive features that make them well-suited to their environment. They possess a unique digestive system with three stomach compartments that enable them to efficiently process fibrous desert vegetation. Their ability to conserve water and withstand extreme temperatures is a result of their specialized physiology.

Camels have a robust skeletal structure designed for endurance and strength. Their long legs and padded feet are adapted for walking long distances on sand. Additionally, camels have a specialized respiratory system that helps them to breathe more efficiently in hot and dry conditions. Their large, oval-shaped nostrils can close to keep out sand and dust.

Understanding these anatomical features is crucial for effective camel husbandry. For instance, recognizing signs of dehydration or heat stress involves knowing how a camel's body responds to environmental stressors. Regular health checks that monitor these physiological aspects can help prevent and address health issues before they become serious.

Historical Significance Of Camels

Camels have a rich historical significance, particularly in the development of trade and transportation across deserts. They were domesticated over 3,000 years ago and played a critical role in the rise of ancient civilizations. Their ability to carry heavy loads over long distances facilitated trade routes such as the Silk Road and the

Incense Route, connecting distant regions and cultures.

Historically, camels were instrumental in the spread of goods, ideas, and cultures. They enabled the movement of people and products between Africa, the Middle East, and Asia. The use of camels in warfare, exploration, and settlement has shaped many historical events and contributed to the growth of ancient empires.

Understanding the historical context of camels helps in appreciating their value and the role they have played in human development. This knowledge can also provide insights into traditional camel husbandry practices and how they have evolved.

Benefits Of Raising Camels

Raising camels offers numerous benefits, both economically and practically. Camels provide a wide range of products, including milk, meat, wool, and hides, which can be utilized for various purposes. Camel milk is known for its high nutritional value and is used in many traditional dishes and beverages. Camel meat is a source of protein, and camel wool is used for clothing and textiles.

In addition to their products, camels serve as an efficient means of transportation, particularly in regions where other vehicles may be impractical. They are capable of carrying heavy loads over long distances and can traverse difficult terrains with ease. This makes them invaluable in both commercial and traditional contexts.

Raising camels also has environmental benefits. Their ability to graze on sparse vegetation without damaging the landscape makes them a sustainable option for pastoral farming. Their dung can be used as a fuel source, reducing dependence on other forms of energy.

By incorporating camels into agricultural practices, farmers can enhance their productivity and sustainability.

CHAPTER TWO

Selecting The Right Camel

Choosing Camel Breeds For Specific Needs

Selecting the right breed of camel is crucial depending on your intended purpose, whether it's for work, milk production, meat, or companionship. There are several breeds to consider, each with distinct characteristics suited to various functions. For instance, the Dromedary (one-humped camel) is ideal for work and transportation in arid regions due to its endurance and adaptability. On the other hand, the Bactrian (two-humped camel) is more suited for colder climates and is often used for its strength in carrying heavy loads.

If your focus is on milk production, breeds such as the Sahiwalor Kharai camels are known for their high milk yield and quality. For meat production, the Mongolian camel offers good body weight and meat quality. Companion camels, often referred to as "pets," are usually selected from breeds with a calm temperament and good trainability, such as the Bactrian or specially bred varieties that are known for their gentle nature.

When selecting a breed, consider the environment and the specific needs of your farm or situation. For example, if you live in a desert area, choose breeds known for their heat tolerance and water efficiency. In contrast, if you're in a temperate region, look for breeds that can handle varying weather conditions. Understanding these factors will help you select the most suitable camel breed for your needs.

Key Characteristics To Look For

When choosing a camel, certain key characteristics will help ensure you pick a healthy and suitable animal. First, examine the camel's body condition. A healthy camel should have a well-rounded body, with visible ribs not overly protruding. The camel's coat should be clean and free from parasites or skin conditions. Additionally, check the eyes; they should be bright and clear, not cloudy or dull, which could indicate illness.

Another important factor is the camel's teeth. Healthy camels should have clean, white teeth without excessive wear or decay. Young camels typically have more robust teeth, while older ones might show signs of wear, which could affect their ability to graze efficiently. They hooves should also be examined; they should be solid and free from cracks or signs of infection.

Behavioral traits are equally important. A good camel should be alert, responsive, and show a generally calm demeanor. Aggressive or overly nervous camels might not be suitable for work or breeding purposes. Observing the camel's interactions with others and its overall demeanor can provide insights into its temperament and suitability for your needs.

Health Indicators In Camels

Evaluating a camel's health is essential before making a purchase. One of the primary indicators is its general appearance. A healthy camel should have a well-proportioned body, with no signs of malnutrition or obesity. Check for symptoms of illness, such as coughing, nasal discharge, or abnormal droppings. These could indicate underlying health issues.

Vaccination and deworming records are important to verify. Ensure the camel is up-to-date on vaccinations and has been regularly deformed. Lack of proper medical care can lead to serious health problems and affect the camel's productivity.

Body temperature can also be a good indicator of health. The normal body temperature for a camel ranges between 99.5 to 102.5 degrees Fahrenheit. A camel with a significantly higher or lower temperature may be suffering from a fever or other health issues. Additionally, check for hydration levels; a well-hydrated camel will have moist mucous membranes and skin that returns to normal quickly when pinched.

Understanding Camel Age And Gender

Knowing the age and gender of a camel is crucial for determining its suitability for specific purposes. Age can be estimated by examining the camel's teeth, as they wear down and change as the animal matures. Young camels (less than three years old) have smoother, well-aligned teeth, while older camels will show more wear and tear.

Gender plays a significant role depending on the intended use. Female camels are preferred for milk production and breeding due to their reproductive capabilities. Male camels, particularly those that are uncastrated, are often used for work and transport due to their strength. Castrated males, known as geldings, can be suitable as companions or for lighter work.

When assessing age and gender, also consider the camel's reproductive health if you're interested in breeding. Females should have regular estrous cycles and males should have good sperm quality. For work or transport camels, ensure they have the strength and stamina appropriate for their age and physical condition.

Tips For Buying Camels

When buying camels, it's essential to purchase from reputable sources. Start by researching breeders or sellers with good reviews and a history of providing healthy animals. Visit the seller's facility if possible to inspect the camels' living conditions and their overall health.

Ask for documentation related to the camel's health history, including vaccination records, deworming schedules, and any previous medical treatments. It's also wise to request a pre-purchase veterinary examination to ensure the camel is in good health and free from diseases.

Negotiate the price based on the camel's age, health status, and purpose. Don't rush into a purchase; take your time to find a camel that fits your needs and budget.

Finally, ensure you understand the care and maintenance requirements for the breed you choose to ensure a successful and rewarding experience with your new camel.

CHAPTER THREE

Camel Housing And Environment

Designing A Camel Pen Or Shelter

Designing a suitable camel pen or shelter is crucial for ensuring the health and well-being of your camels. Start by selecting a location that is well-drained and away from potential hazards such as flooding or strong winds. The shelter should provide adequate protection from extreme weather conditions while allowing for easy management and access.

A typical camel pen should be spacious enough to allow each camel to move around comfortably. Generally, a space of at least 200-250 square feet per camel is recommended. The shelter should include separate areas for feeding, resting, and exercise. Constructing the pen using sturdy materials such as timber, metal, or masonry will help ensure durability and safety. Ensure the walls are high enough to prevent the camels from jumping over and that the enclosure is secure to prevent escapes.

For the roof, consider materials that offer good insulation and protection from both heat and cold. A thatched roof or corrugated metal sheets with insulation layers can be effective in maintaining a stable internal temperature. Additionally, design the shelter to include sliding doors or gates for easy entry and exit, and consider adding covered areas to protect the camels from sun exposure and rain.

Necessary Equipment And Supplies

Equipping your camel pen with the right tools and supplies is essential for daily management and ensuring the camel's comfort.

Start with basic supplies such as feed troughs, water containers, and bedding material. Feed troughs should be designed to minimize waste and facilitate easy cleaning. For water containers, opt for large, durable troughs that can hold enough water for the camels to drink throughout the day.

Bedding is an important aspect of camel housing as it helps maintain cleanliness and comfort. Straw, hay, or sawdust are commonly used bedding materials that are easy to manage and dispose of. Make sure to provide a sufficient amount of bedding to absorb moisture and reduce odors. Regularly replace soiled bedding to maintain hygiene.

Other essential equipment includes grooming tools, medical supplies, and waste management tools. Grooming tools like brushes and combs help in maintaining the camels' coats, while a first-aid kit with basic medications and bandages is necessary for treating minor injuries.

Waste management tools such as shovels and manure forks will assist in keeping the pen clean and odor-free.

Maintaining Cleanliness And Hygiene

Maintaining cleanliness and hygiene in the camel pen is vital for preventing disease and ensuring a healthy environment. Begin with a daily cleaning routine that includes removing soiled bedding, manure, and any spilled feed. Regularly wash and disinfect feed troughs and water containers to prevent contamination.

Implement a waste disposal system to manage manure effectively. Manure should be collected daily and disposed of properly to avoid attracting pests and creating unpleasant odors. Consider composting the manure, which can be a beneficial by-product for use as fertilizer in gardens or fields.

In addition to daily cleaning, schedule regular deep-cleaning sessions where the entire pen is scrubbed and disinfected. This helps to eliminate bacteria and parasites that may thrive in dirty conditions. Use non-toxic disinfectants and ensure that all cleaning agents are thoroughly rinsed away to avoid harming the camels.

Importance Of Space And Ventilation

Space and ventilation are critical factors in maintaining a healthy environment for camels. Adequate space allows camels to move freely, reducing stress and the risk of injuries. Overcrowding can lead to aggressive behavior and health issues, so it is essential to provide enough space per camel.

Ventilation is equally important as it helps to maintain fresh air circulation and prevent the buildup of harmful gases such as ammonia from urine and manure. Ensure that the shelter has windows or vents that can be opened to allow for natural airflow. If the climate is particularly humid or hot, consider installing fans or ventilation systems to enhance air movement and regulate temperature.

The design of the pen should also account for the direction of prevailing winds to avoid drafts that can lead to respiratory problems. Proper ventilation combined with adequate space will create a more comfortable and healthier living environment for the camels.

Protecting Camels From Extreme Weather

Protecting camels from extreme weather conditions is essential for their well-being and productivity. In hot climates, provide ample shade and access to cool water to prevent heat stress.

Shade structures such as large roof overhangs or trees can help reduce exposure to direct sunlight. Ensure that water containers are always filled with clean, cool water.

In cold climates, it is important to provide shelter that is well-insulated and protected from wind and rain. Insulated roofs and windbreaks can help keep the shelter warm and dry. During harsh weather, monitor the camels closely for signs of cold stress and provide additional bedding for warmth.

Additionally, consider using windbreaks or barriers around the shelter to protect against strong winds and snow. Regularly inspect the shelter for any damage and make necessary repairs to maintain its effectiveness in protecting the camels from extreme weather conditions.

CHAPTER FOUR

Feeding And Nutrition

Essential Nutrients For Camels

Camels often referred to as the "ships of the desert," have unique nutritional needs that are crucial for their health and productivity. Understanding these essential nutrients is vital for anyone involved in camel husbandry. The primary nutrients camels require include carbohydrates, proteins, fats, vitamins, and minerals. Carbohydrates provide energy, which is essential for camels to perform their daily activities and withstand the harsh desert environment. Sources of carbohydrates include hay, grains, and some types of forage.

Proteins are crucial for growth, reproduction, and overall health. Camels need high-quality protein sources such as legumes, alfalfa, and soybean meal. Proteins help in tissue repair and maintenance, which is especially important for young camels and lactating females. Fats are also important as they supply energy and aid in the absorption of fat-soluble vitamins. Common fat sources include vegetable oils and certain grains.

Vitamins and minerals play a key role in maintaining various bodily functions. For instance, vitamin A is important for vision and skin health, while vitamin D aids in calcium absorption. Minerals such as calcium, phosphorus, and salt are critical for bone development and overall health. Providing a balanced mineral supplement can help ensure that camels receive these essential nutrients in the right amounts.

Types Of Camel Feed And Supplements

Choosing the right type of feed and supplements is crucial for maintaining camel health. Camels can be fed a combination of roughage and concentrate feeds. Roughage includes grasses, hay, and silage, which provide the necessary fiber for proper digestion. Forage should be of good quality, free from mold or contaminants, and properly stored to prevent spoilage.

Concentrates, such as grains and pellets, are energy-dense and should complement the roughage. Common concentrates include barley, oats, and corn. These should be fed in moderation to avoid digestive issues and obesity. Additionally, commercial camel feed pellets are available, which are specially formulated to meet the nutritional needs of camels and often include added vitamins and minerals.

Supplements can be beneficial to address specific nutritional gaps. Mineral blocks and vitamin supplements can help in maintaining overall health. For camels that are pregnant or lactating, high-quality protein and calcium supplements may be necessary to support the increased nutritional demands. Always follow the manufacturer's guidelines for the correct dosage and administration of supplements.

Feeding Schedule And Quantity

Establishing a proper feeding schedule and determining the right quantity of feed is essential for optimal camel health. Camels should be fed at regular intervals to ensure they receive a steady supply of nutrients. Typically, camels are fed twice a day, with the option of additional feeding if needed, depending on their age, activity level, and overall health.

For adult camels, the daily feed quantity can vary, but a general guideline is about 2-3% of their body weight in dry matter. For example, a camel weighing 600 kg should receive approximately 12-18 kg of dry feed per day. This can be adjusted based on the camel's condition, workload, and reproductive status.

Young camels and lactating females have different nutritional requirements. Calves and weanlings may require more frequent feeding to support growth while lactating females need additional energy and protein to produce milk. It is important to monitor the camel's body condition and adjust feed quantities accordingly to prevent under or overfeeding.

Common Nutritional Issues And Solutions

Nutritional deficiencies can lead to a range of health issues in camels. Common problems include weight loss, poor coat condition, and reproductive issues.

To address these, it's crucial to identify the specific nutrient that may be lacking. For instance, a camel with a dull coat might be suffering from a lack of essential fatty acids or vitamins.

One common issue is metabolic disorders caused by imbalanced feeding. For example, overfeeding on concentrates can lead to obesity and digestive disturbances. To prevent this, ensure a balanced diet with appropriate roughage and limit concentrate intake to recommended levels. Providing a variety of feed sources can help in achieving a balanced diet and preventing deficiencies.

If you observe any signs of nutritional deficiencies, consult with a veterinarian or animal nutritionist to adjust the diet accordingly. Regular health checks and feed analysis can help in identifying and addressing potential issues before they become severe.

Water Requirements And Management

Water is critical for camel health and performance. Camels are known for their ability to survive with minimal water, but they still require access to clean and fresh water regularly. On average, a camel needs about 5-10 liters of water per 100 kg of body weight per day, although this can vary depending on environmental conditions and activity levels.

Proper water management involves ensuring that camels have access to water at all times. In hot climates, camels may need to drink more frequently, and water sources should be checked regularly for cleanliness and availability. Providing shaded areas near water sources can also encourage camels to drink more.

In addition to providing adequate water, it's important to monitor water quality. Contaminated water can lead to health issues, so ensure that water containers are clean and free from debris. In areas with limited water supply, consider strategies for water conservation and management, such as using water-saving techniques and ensuring efficient use of available resources.

CHAPTER FIVE

Camel Health And Veterinary Care

Common Health Issues In Camels

Camels, like any other livestock, are susceptible to a range of health issues that can affect their well-being and productivity. One common issue is foot rot, which is a bacterial infection affecting the hooves. This condition often results from poor hygiene or wet, muddy conditions. To manage foot rot, ensure that camels are kept in clean, dry environments and inspect their hooves regularly for signs of infection. Prompt treatment with antibiotics and proper hoof care are essential to prevent severe complications.

Another prevalent health problem is digestive disorders such as colic or bloat. Camels are prone to these issues due to their unique digestive system, which requires a balanced diet and proper feeding practices. To avoid digestive problems, provide camels with a diet rich in fiber and avoid sudden changes in feed. Regularly check for symptoms like discomfort, distended abdomen, or changes in feces, and consult a veterinarian for appropriate treatments, which might include medications or dietary adjustments.

Respiratory infections are also a concern, especially in camels exposed to extreme weather conditions or poor ventilation. Signs of respiratory issues include coughing, nasal discharge, and labored breathing. To prevent these infections, maintain good barn ventilation, avoid overcrowding, and provide proper shelter from harsh weather. If a camel shows signs of respiratory illness, a veterinarian may recommend antibiotics or other treatments to alleviate symptoms and prevent the spread of infection.

Routine Health Checks And Vaccinations

Routine health checks and vaccinations are crucial for maintaining the overall health of camels and preventing disease outbreaks. A standard health check involves a thorough examination of the camel's physical condition, including checking their weight, body condition, and vital signs such as temperature, pulse, and respiration rate. Regularly monitoring these parameters helps in early detection of potential health issues. Health checks should be conducted at least twice a year or more frequently if any health concerns arise.

Vaccinations play a significant role in preventing infectious diseases in camels. Common vaccines include those for brucellosis, foot-and-mouth disease, and camel pox. These vaccines are typically administered based on a schedule recommended by a veterinarian. For instance, the brucellosis vaccine is usually given to young camels around six months of age, with booster shots administered as needed. Keeping an up-to-date vaccination record and adhering to the vaccination schedule is crucial for ensuring your camels remain protected against preventable diseases.

In addition to vaccinations, routine deworming is necessary to control internal parasites that can affect camels' health and productivity. Use dewormers as prescribed by a veterinarian, and follow a deworming schedule to manage parasite loads effectively. Regular fecal examinations can help determine the need for deworming treatments and ensure that your camels are free of parasites.

Recognizing Signs Of Illness

Early detection of illness in camels can significantly impact their recovery and overall health. Behavioral changes such as lethargy, loss of appetite, or isolation from the herd can be early indicators of illness. Camels that are not eating or drinking normally should be monitored closely, as these behaviors often precede more serious health issues. Changes in behavior should prompt a thorough examination and consultation with a veterinarian.

Physical symptoms are also crucial in identifying health problems. Look out for abnormalities such as swelling, discharge, or changes in the camel's coat condition. For instance, a swollen joint might indicate an infection or injury, while abnormal discharge could point to respiratory or reproductive issues. Monitoring the camel's body temperature is also important; an elevated temperature can be a sign of fever or infection.

In addition to physical and behavioral signs, changes in droppings can indicate health problems. Diarrhea, constipation, or irregular droppings can signal digestive issues or parasitic infections.

Regularly inspecting the camel's manure and seeking veterinary advice when changes are observed can help manage these conditions effectively.

Emergency First Aid For Camels

Being prepared for emergencies is essential in camel husbandry, as quick and effective first aid can make a significant difference in the outcome of a health crisis. For minor cuts and wounds, clean the area with an antiseptic solution to prevent infection. Apply an appropriate wound dressing and monitor the injury for signs of healing or complications. If the wound is deep or bleeding heavily, seek veterinary assistance immediately.

In cases of heatstroke, which can occur in hot climates, move the camel to a cooler area and provide plenty of water. Use cool, wet cloths to lower the camel's body temperature gradually. Avoid using ice-cold water or placing the camel in a cold environment suddenly, as this can cause shock. Monitoring the camel closely and seeking veterinary help is crucial if symptoms persist.

For dystopia (difficult calving), ensure the camel is in a clean, safe environment and assist with the birth if necessary. Use clean hands and equipment to avoid introducing infections. If complications arise, such as prolonged labor or signs of distress in the camel, contact a veterinarian for guidance and possible intervention.

Finding And Working With A Veterinarian

Finding a knowledgeable and experienced veterinarian is crucial for effective camel health management. Start by seeking recommendations from other camel owners or local agricultural extension services. Look for a veterinarian who has experience with camels and understands their specific health needs. Ensure that the veterinarian has a good reputation and is willing to provide comprehensive care.

Once you have selected a veterinarian, establish a clear line of communication and discuss your camel husbandry practices and health concerns. Regular visits and open communication will help the veterinarian understand your camel's specific needs and tailor their care accordingly. Keep detailed records of health checks, vaccinations, and any treatments provided, and share this information with your veterinarian during consultations.

Building a good relationship with your veterinarian involves being proactive about your camel's health and seeking advice on preventive measures and emergency care. Follow the veterinarian's recommendations for vaccinations, deworming, and other health management practices.

CHAPTER SIX

Basics Of Camel Reproduction

Camel reproduction is an intricate process that requires understanding both the biological and practical aspects. Camel breeding typically starts with recognizing the right time for mating, which is influenced by the camel's reproductive cycle. Female camels, or cows, come into estrus or heat, every 14 to 30 days, and this period lasts for about 2 to 6 days. During this time, they are most receptive to mating.

To ensure successful reproduction, it is essential to monitor the signs of estrus in female camels. These signs include restlessness, increased vocalization, and a willingness to mount or be mounted. Males, or bulls, also play a crucial role in this process, and their readiness can be determined by their mating behavior, which includes exhibiting dominance and interest in females.

Effective camel reproduction management involves selecting breeding pairs based on genetic qualities and health. Choose healthy, strong camels with desirable traits such as good milk production, strong constitution, and resistance to common diseases. Regular veterinary check-ups and vaccinations are also necessary to maintain the overall health of both males and females, ensuring a higher success rate in breeding.

Breeding Techniques And Practices

Breeding camels requires a combination of traditional knowledge and modern practices. The most common method of camel breeding is natural mating, where male and female camel are allowed to mate freely.

This method is straightforward but requires careful observation to ensure that mating occurs successfully and that the camels are not harmed during the process.

Artificial insemination (AI) is another breeding technique used to improve genetic quality and manage breeding more effectively. AI involves collecting semen from a male camel and then introducing it into the reproductive tract of a female camel during her estrus period. This method requires specialized equipment and expertise but offers greater control over breeding and helps in the selection of superior genetic traits.

In both natural mating and artificial insemination, it is crucial to keep accurate records of breeding dates, camel health, and any signs of reproductive issues. This information helps in planning future breeding and in managing the overall reproductive health of the herd.

Using these records, breeders can identify the best times for mating and can track the genetic progress of the herd over time.

Gestation And Birth Process

The gestation period for camels typically lasts about 13 months, which is longer than most other livestock. During this time, the pregnant female camel, or cow, requires special care to ensure the health of both herself and the developing fetus. Nutritional needs are particularly important; a balanced diet rich in proteins, vitamins, and minerals supports a healthy pregnancy. Fresh water and ample forage should be available at all times.

As the birth approaches, signs of impending labor include increased restlessness, nesting behavior, and the appearance of a full udder. Camels usually give birth standing up, and the process can take several hours.

Providing a clean, dry, and quiet environment for the birth is essential to minimize stress on the mother and the newborn calf.

After birth, the newborn camel, called a cria, should be monitored closely to ensure it starts nursing within a few hours. Colostrum, the first milk produced by the mother, is crucial for the cria's immune system. Ensure the mother and baby have a warm and secure space to bond and recover. Regular monitoring and veterinary care are essential during the early days to address any issues that may arise and to support the health of both the mother and the newborn.

Care For Pregnant And Newborn Camels

Caring for pregnant camels involves more than just providing food and water; it includes regular health check-ups and monitoring for any signs of complications. Pregnant camels should be provided with a comfortable and safe environment to reduce stress and prevent injuries. It's also important to maintain a regular vaccination and deworming schedule to protect the health of both the mother and the unborn calf.

For newborn camels, immediate care includes ensuring they are warm and dry, as they are vulnerable to temperature fluctuations. Newborns should be checked to ensure they are nursing properly and receiving adequate colostrum. The mother's udder should be monitored to prevent any issues like mastitis, which can affect milk production and the newborn's health.

Proper handling and socialization are crucial during the first few weeks. Gentle interactions help the newborn camel become accustomed to human contact, which is important for its future health and management.

Regular veterinary check-ups are necessary to monitor the growth and development of the cria and to address any potential health issues early on.

Managing Camel Herds For Optimal Reproduction

Effective herd management is key to optimizing reproduction and ensuring a healthy, productive herd. Regular health checks, vaccinations, and deworming schedules should be strictly followed. Implementing a systematic breeding program helps in planning the breeding cycles and managing the genetic quality of the herd.

Record-keeping plays a vital role in managing camel herds. Detailed records of breeding dates, health issues, and offspring performance help in making informed decisions about future breeding and herd management.

This data also assists in tracking the genetic progress of the herd and identifying any recurring issues that may need addressing.

Providing a suitable environment for camels is also crucial. This includes ample grazing space, clean water, and shelter from extreme weather conditions. Stress reduction through proper handling and socialization practices enhances reproductive success and overall herd health. By following these practices, camel breeders can achieve a balanced and productive herd, ensuring both high reproductive rates and healthy animals.

CHAPTER SEVEN

Training And Handling Camels

Basic Training Techniques For Camels

Training camels involves a blend of patience, consistency, and positive reinforcement. Begin with establishing a routine that camels can predict and rely on. Start by spending time with your camel every day, building a bond through gentle interaction. Approach the camel calmly and use a soft voice to make it comfortable. Regularly handling and touching your camel will help it get used to human presence and touch.

Introduce basic commands slowly. For instance, use simple phrases like "Come" or "Stand" while gesturing with your hand.

Consistently repeat these commands so that the camel can associate the verbal command with the action. Reinforce correct behavior with rewards such as treats or affection. Consistency is key, as camels need to understand that the commands and their rewards are linked. Always end training sessions on a positive note to keep the camel eager to participate in future sessions.

Building Trust And Handling Skills

Building trust with a camel is fundamental for effective training and handling. Start by creating a positive environment where the camel feels secure. Spend time feeding and grooming your camel to establish a bond. Gentle brushing, talking, and the occasional treat can help the camel associate you with positive experiences. Avoid sudden movements or loud noises, as these can startle the camel and hinder trust-building.

Handling skills are developed through repeated, calm interactions. Begin with leading the camel on a lead rope in a controlled area, guiding it with gentle tugs rather than forceful pulls. Practice walking alongside the camel, and gradually introduce activities like haltering and tying. Always ensure the camel is comfortable and relaxed during these activities. Use positive reinforcement to reward calm behavior and progress, helping the camel feels more secure and cooperative.

Training For Specific Tasks (E.G., Riding, Packing)

Training camels for specific tasks, such as riding or packing, requires clear, gradual steps. For riding, start by getting the camel accustomed to having a saddle placed on its back. Allow the camel to get used to the saddle without weight first, rewarding it for staying calm.

Once the camel is comfortable with the saddle, begin by gently mounting it, starting with short periods and gradually increasing as the camel becomes more accustomed to the process.

For packing, introduce the camel to the saddle or pack frame without weight initially. Let the camel get used to the sensation of the pack on its back. Begin with light loads, and gradually increase the weight as the camel adjusts. Ensure that the pack is evenly distributed and does not cause discomfort. Consistently reward the camel for accepting and carrying the pack, making the experience as positive as possible. Regular practice and patience will help the camel adapt to these specific tasks effectively.

Addressing Behavioral Issues

Addressing behavioral issues in camels requires observation and understanding of the root causes. Common issues include aggression, fearfulness, or resistance to training. Start by identifying triggers for these behaviors. For instance, if a camel becomes aggressive, it may be due to fear or discomfort. Evaluate the camel's environment, health, and handling methods to determine possible causes.

Implement a systematic approach to address these issues. For aggressive behavior, ensure that handling is calm and non-threatening. Gradually desensitize the camel to situations that trigger aggression by introducing them in a controlled, positive manner. For fearfulness, increase the camel's exposure to the source of fear gradually, paired with rewards for calm behavior. Consistent, positive reinforcement is crucial in modifying undesirable behaviors. Avoid punishment, as it can exacerbate fear and aggression.

Safety Tips For Handling Camels

Handling camels safely requires awareness of their size and strength. Always approach camels from the side, where they can see you coming, to avoid startling them. Use appropriate equipment such as a halter and lead rope to maintain control, and ensure that the equipment fits well and is comfortable for the camel. When leading a camel, stay aware of its body language to gauge its mood and prevent sudden reactions.

Avoid handling camels when they are agitated or in discomfort. If a camel exhibits signs of distress or aggression, give it space and allow it to calm down before resuming handling. Ensure that you are physically prepared for the camel's strength, and enlist help if needed to manage larger or more unruly camels. Training in a safe, enclosed area helps prevent accidents. Regularly check the condition of all handling equipment to ensure it is in good working order.

CHAPTER EIGHT

Camel Products And Uses

Milk Production And Processing

Camel milk is a valuable resource with distinct nutritional benefits, making it an essential product in camel husbandry. The production process begins with ensuring the health and well-being of the camels, as their milk yield is directly affected by their diet and living conditions. Begin by providing a balanced diet rich in nutrients, including high-quality forage, grains, and freshwater. Regular veterinary check-ups and maintaining a clean environment are crucial for optimal milk production.

When milking, choose a consistent schedule, preferably twice a day. Clean the udder and teats thoroughly before milking to prevent contamination. Use clean, sanitized equipment to collect the milk.

Start by gently massaging the udder to stimulate milk flow, and then carefully squeeze and pull the teats in a rhythmic motion. Once collected, the milk should be cooled immediately to preserve its freshness.

Processing camel milk involves pasteurization to ensure safety and extend shelf life. Heat the milk to 72°C (161°F) for 15 seconds, then quickly cool it down. Store pasteurized milk in clean, airtight containers in a refrigerator. Camel milk can also be turned into various products like yogurt and cheese. For yogurt, add a starter culture to the milk, incubate it at 40-45°C (104-113°F) for several hours, and then refrigerate. For cheese, use a coagulant to curdle the milk, drain the curds, and press them into molds.

Meat And Wool Utilization

Camel meat is a nutritious and highly valued food source, particularly in arid regions where camels are commonly raised. The process of utilizing camel meat starts with proper slaughtering techniques. Ensure that the camel is well-fed and hydrated before slaughter. The meat should be processed in a clean and hygienic environment to prevent contamination. After slaughter, the carcass is skinned and the meat is divided into various cuts, such as steaks, roasts, and ground meat.

Camel wool, on the other hand, is a significant by-product, known for its warmth and durability. Wool harvesting is done annually, usually in the spring when camels naturally shed their winter coats. Use a comb or shearing equipment to collect the wool. Handle the wool gently to avoid damage. After harvesting, the wool should be cleaned to remove dirt and grease. This involves washing it in lukewarm water with a mild detergent, followed by rinsing and drying.

Both camel meat and wool can be processed and utilized in different ways. Meat can be preserved through methods like freezing, drying, or curing. Camel wool can be spun into yarn for knitting or weaving, and crafted into various products such as blankets, clothing, and carpets. Proper processing and handling ensure that both meat and wool retain their quality and value.

Leather And Craft Products

Camel leather is renowned for its strength and flexibility, making it an excellent material for various craft products. The first step in utilizing camel leather is the tanning process, which converts raw hides into durable leather. Begin by cleaning the hides thoroughly to remove any flesh or fat.

Next, soak the hides in a solution of water and tannins, which can be derived from plant sources or synthetic chemicals. This process stabilizes the collagen in the hides and makes them resistant to decay.

Once tanned, the leather is dyed and treated to enhance its appearance and functionality. Various dyes and finishes can be applied, depending on the desired outcome. For crafting, camel leather can be cut and shaped into products like bags, belts, shoes, and saddles. Use specialized tools for cutting and stitching to ensure precision and durability. Leathercrafting techniques include hand-stitching, embossing, and tooling to create intricate designs and patterns.

Proper maintenance of camel leather products is essential for their longevity. Store leather items in a cool, dry place and clean them regularly using appropriate leather care products.

Avoid exposing leather to excessive moisture or direct sunlight, as this can cause damage. Regular conditioning with leather cream helps keep the leather supple and prevents it from drying out.

Economic Value Of Camel Products

The economic value of camel products is significant, particularly in regions where camels are a primary resource. Camel milk, meat, wool, and leather contribute to both subsistence and commercial economies. Camel milk is often sold fresh or processed into value-added products like cheese and yogurt, providing a steady income stream for camel herders. The market demand for camel milk is growing due to its health benefits and unique properties.

Camel meat is a high-value commodity in many cultures, and its demand is often seasonal, peaking during festivals and special occasions. The meat's nutritional benefits and flavor make it a popular choice, adding to its economic value. Camel wool and leather also hold considerable economic importance. Wool is used in traditional textiles and modern fashion, while leather products are sought after for their quality and durability.

By understanding and leveraging the economic value of camel products, camel herders can enhance their livelihoods. Diversifying product offerings and exploring new markets can increase profitability. Additionally, investing in quality processing and marketing strategies can further boost the economic impact of camel husbandry.

Marketing And Selling Camel Products

Marketing and selling camel products effectively require a strategic approach to reach potential buyers and maximize sales.

Start by identifying your target market, which could include local consumers, specialty food stores, or international buyers interested in exotic or health-conscious products. Develop a strong brand identity that highlights the unique qualities of camel products, such as their nutritional benefits or artisanal craftsmanship.

Create a marketing plan that includes online and offline strategies. Establish an online presence through a website or social media platforms to showcase your products and connect with customers. Use high-quality images and detailed descriptions to attract interest. Participate in local markets, fairs, or trade shows to reach customers directly and build relationships with buyers.

Pricing strategy is crucial for successful sales. Consider factors such as production costs, market demand, and competition when setting prices. Offering promotions or discounts can attract customers and encourage bulk purchases. Additionally, providing excellent customer service and maintaining product quality will help build a loyal customer base and foster positive word-of-mouth recommendations.

By implementing effective marketing and selling strategies, camel herders can enhance the visibility and appeal of their products, leading to increased sales and greater economic success.

CHAPTER NINE

Challenges And Solutions In Camel Husbandry

Common Challenges Faced By Camel Owners

Camel husbandry, while rewarding, comes with its own set of challenges. One of the primary difficulties camel owners face is managing the health of these large animals. Camels are susceptible to various diseases, including zoonotic diseases that can also affect humans, such as brucellosis and tuberculosis. Ensuring camels are properly vaccinated and monitored regularly for signs of illness is crucial. Additionally, camels can suffer from parasitic infections, such as internal worms and external pests like ticks and lice. These issues can be managed through regular deworming and pest control measures, but detecting and addressing them promptly is key to preventing severe health problems.

Another challenge is the camel's dietary needs. Camels have specific nutritional requirements that differ from other livestock. They thrive on a diet that includes a variety of forage, such as hay, and access to fresh water. A common issue is providing adequate and balanced nutrition, especially in areas where natural forage is scarce. This can be addressed by supplementing their diet with appropriate feed additives and ensuring they have constant access to clean water. Additionally, camel owners often struggle with the cost and availability of quality feed, particularly in remote areas.

Camels are also known for their unique social behavior, which can pose challenges in a husbandry setting. For instance, camels are hierarchical animals, and conflicts can arise if they are not properly managed. Dominance disputes and aggressive behavior among camels can lead to injuries and stress. Effective management strategies include providing adequate space, ensuring a clear social hierarchy, and minimizing overcrowding. Implementing regular training and handling routines can help mitigate these issues and ensure a harmonious environment for the camels.

Solutions And Best Practices

Addressing the challenges of camel husbandry involves implementing best practices that ensure the health, nutrition, and well-being of the camels. One effective solution for health management is establishing a comprehensive veterinary care routine. This includes regular health check-ups, vaccinations, and preventative treatments for parasites.

It's beneficial to work closely with a veterinarian who has experience with camels to develop a tailored health management plan.

For nutritional management, creating a balanced diet plan is essential. Camel owners should focus on providing a variety of high-quality forages and supplements. Adding nutritional supplements that are specifically designed for camels can help meet their dietary requirements. Additionally, investing in feed storage solutions that prevent spoilage and contamination can ensure that the camels receive consistent and safe nutrition.

In terms of social management, implementing a structured environment can reduce behavioral issues. Providing ample space and shelter for camels to retreat from one another if needed can help prevent aggression. Regular interaction and training sessions can help establish and reinforce a positive social structure within the herd.

Training camels to respond to commands and handling can also facilitate easier management and reduce the risk of injuries.

Adapting To Climate And Environmental Changes

Camels are well-suited to arid environments, but climate and environmental changes can still pose significant challenges. For instance, extreme temperatures and varying levels of precipitation can impact the availability of forage and water sources. To adapt to these changes, camel owners should implement strategies to manage their resources effectively.

During periods of drought, it's crucial to have a reliable source of water and to explore options for water conservation and storage. Implementing rainwater harvesting systems and providing supplementary water sources can help ensure that camels remain hydrated. Additionally, providing shade and shelter to protect camels from extreme heat can help reduce stress and prevent heat-related illnesses.

In areas where environmental conditions are changing, such as increased rainfall leading to overgrown vegetation, managing forage quality becomes important. Regularly assessing and adjusting the diet based on the available forage and supplementing with commercial feed if necessary can help maintain the health of the camels. Adapting to climate changes also involves monitoring and responding to shifts in the camels' health and behavior, which can indicate changes in environmental conditions.

Resources For Ongoing Support

Staying informed and supported is vital for successful camel husbandry. Numerous resources are available to camel owners to help them address challenges and improve their practices. Veterinary associations and livestock extension services often provide valuable information and support for camel health management. Online forums and networks for camel owners can also offer practical advice and solutions from experienced peers.

Educational resources, such as workshops, webinars, and publications focused on camel husbandry, can provide up-to-date information on best practices and emerging trends. Joining local or regional camel associations can offer additional support through networking opportunities, shared resources, and access to specialized training.

Furthermore, seeking advice from agricultural extension officers and consultants who specialize in camel husbandry can provide tailored guidance and solutions based on specific regional challenges. Investing time in these resources can help camel owners stay informed and improve their husbandry practices effectively.

CHAPTER TEN

Legal And Ethical Considerations

Understanding Local Regulations And Laws

In camel husbandry, adhering to local regulations and laws is crucial for ensuring a successful and compliant operation. The first step in understanding these regulations is to research the specific laws and requirements related to camel care in your region. This involves contacting local agricultural departments, veterinary services, and animal welfare organizations to obtain detailed information on licensing, zoning, and animal welfare standards. In some areas, you may need permits for keeping camels, especially if you are running a commercial operation.

Next, ensure that you comply with health and safety regulations. This includes understanding quarantine procedures, vaccination requirements, and disease prevention measures that apply to camels. Local regulations may also dictate the construction standards for shelters and fencing, as well as the protocols for waste management. Regularly reviewing and updating your knowledge of these regulations helps prevent legal issues and ensures that your husbandry practices remain in line with current standards.

It's also beneficial to join local agricultural or camel husbandry associations. These groups often provide resources and updates on legal changes and best practices. They can offer guidance and support, helping you stay informed about any new regulations that might affect your operation.

Ethical Treatment Of Camels

The ethical treatment of camels is a fundamental aspect of camel husbandry. Ensuring the well-being of your camels involves providing proper nutrition, medical care, and living conditions. Start by creating a balanced diet tailored to the specific needs of camels, which includes high-quality forage, grains, and freshwater. Regular health check-ups with a veterinarian are essential to monitor the camels' health and address any issues promptly.

Additionally, ensure that the living environment for your camels is spacious, clean, and safe. Proper ventilation, shelter from extreme weather, and ample space for exercise are crucial for their well-being. Regular cleaning of their living area helps prevent the spread of diseases and parasites.

Training and handling should be done with care and respect, avoiding any practices that could cause stress or harm to the animals. Positive reinforcement techniques are recommended for training, ensuring that interactions are based on trust and respect rather than fear.

Record-Keeping And Documentation

Effective record-keeping is vital for managing a successful camel-husbandry operation. Start by maintaining comprehensive records of each camel, including its health history, vaccinations, and breeding details. Use a dedicated logbook or digital management system to track these records, ensuring that they are up-to-date and easily accessible.

Document all aspects of your camel husbandry practices, such as feed consumption, growth rates, and any medical treatments administered. This information is invaluable for monitoring the health and productivity of your camels, identifying trends, and making informed decisions.

Regularly review and audit your records to ensure accuracy and completeness. This practice not only helps in maintaining high standards of care but also supports compliance with legal and regulatory requirements. Additionally, having well-organized documentation can be beneficial in case of inspections or when seeking advice from veterinarians or other professionals.

Responsible Breeding Practices

Responsible breeding practices are essential for maintaining the health and quality of your camel herd. Begin by selecting a breeding stock that exhibits desirable traits and is free from genetic disorders. Perform thorough health screenings and genetic testing to ensure that the camels you plan to breed are in optimal health and do not carry heritable diseases.

Develop a breeding plan that includes guidelines for mating, gestation, and post-birth care. Monitor pregnant camels closely, providing them with appropriate nutrition and veterinary care throughout their pregnancy. Prepare for the birthing process by setting up a clean and comfortable space for delivery, and be ready to assist if needed.

Post-birth, ensure that both the dam and the newborn camel receive proper care, including immediate veterinary checks, vaccinations, and nutrition. Implementing a responsible breeding program helps improve the overall health and productivity of your camel herd while promoting ethical practices.

Engaging With Camel Husbandry Communities

Engaging with camel husbandry communities can provide valuable support and knowledge for both beginners and experienced keepers. Start by joining local or online camel husbandry groups, forums, and associations. These communities offer opportunities to share experiences, seek advice, and stay informed about industry trends and best practices.

Participate in community events such as workshops, seminars, and trade shows to network with other camel keepers and industry professionals. These events often provide hands-on demonstrations and practical insights that can enhance your husbandry skills.

Additionally, contributing to community discussions and sharing your own experiences can help build a supportive network and foster collaborative relationships. Engaging with camel husbandry communities not only expands your knowledge base but also connects you with a network of like-minded individuals who share your passion for camel care.

Frequently Asked Question And Nswers.

What are the primary benefits of raising camels?

Camels are valuable for their milk, meat, hides, and pack animals. They are also highly adaptable to arid environments and can serve as a source of income and livelihood for pastoral communities.

What do camels eat?

Camels are herbivores and primarily graze on grasses, shrubs, and leaves. They can survive on low-quality forage and have a remarkable ability to extract moisture from their food.

How much water do camels need?

Camels can survive without water for several days, but they typically drink large amounts when available, up to 40 liters (10 gallons) at once.

Their water needs vary based on their diet, climate, and activity level.

How often should camels be fed?

Camels should be fed twice a day, with access to forage throughout the day. Their diet should be balanced with adequate roughage and, if necessary, supplemented with concentrates.

What are common diseases in camels?

Common diseases include camelpox, brucellosis, foot-and-mouth disease, and internal parasites. Regular veterinary care and vaccinations are essential for preventing and managing these diseases.

How can I prevent and manage camel parasites?

Regular deworming and maintaining good hygiene in the camel's living area are key. Consult a veterinarian for appropriate treatments and preventive measures.

What is the gestation period for camels?

The gestation period for camels is approximately 13 months (390 to 410 days). Female camels typically give birth to one calf, although twins can occur.

How can I care for a newborn camel calf?

Ensure the calf receives colostrum (first milk) within the first few hours after birth. Provide a clean, dry environment and monitor the calf's health closely. Ensure it is nursing regularly and gaining weight.

What is the best way to house camels?

Camels need a shelter that protects them from extreme weather conditions, such as excessive heat, cold, or rain. They can be housed in simple structures or open pens with adequate space.

How often should camels be groomed?

Camels should be groomed regularly to maintain their coat and check for signs of parasites or injuries.

Grooming also helps in building a bond between the camel and the handler.

How do I handle aggressive or untrained camels?

Safety is paramount. Use proper training techniques and patience, and, if necessary, seek assistance from a professional camel handler or trainer to manage aggressive behavior.

What are the signs of illness in camels?

Signs of illness may include loss of appetite, lethargy, coughing, nasal discharge, diarrhea, or abnormal behavior. Prompt veterinary attention is essential for proper diagnosis and treatment.

How can I improve camel breeding?

Select healthy breeding stock with desirable traits. Provide optimal nutrition, manage stress levels, and ensure proper veterinary care to improve breeding outcomes.

What is the milking process for camels?

Camels are typically milked twice a day, morning and evening. Use clean equipment and ensure proper hygiene. Regular milking helps maintain milk production and avoid discomfort for the camel.

How do I train a camel for riding or packing?

Start with basic handling and desensitization. Gradually introduce the camel to riding or packing equipment, and use positive reinforcement techniques. Consistent training and patience are key.

What are the common uses of camel hides?

Camel hides are used for making leather products such as shoes, belts, bags, and traditional clothing. The hides are durable and provide a valuable resource for various products.

How can I manage camel waste?

Camel waste can be composted and used as fertilizer. It can also be collected and properly disposed of to maintain a clean and healthy environment.

What are the best practices for camel breeding management?

Implement good nutrition, monitor health regularly, manage mating schedules, and provide a clean and stress-free environment for breeding.

How do camels adapt to extreme temperatures?

Camels have physiological adaptations such as a specialized coat that insulates against both heat and cold, and the ability to regulate their body temperature efficiently. They can also tolerate dehydration and extreme temperatures better than many other animals.

What are the legal requirements for camel husbandry in different regions?

Legal requirements vary by region and may include permits for keeping camels, health and welfare standards, and regulations regarding animal transport and trade. It's important to check local regulations and ensure compliance.

CONCLUSION

Camel husbandry represents a crucial aspect of agriculture and pastoralism in arid and semi-arid regions, where camels are invaluable assets due to their remarkable adaptability to harsh environments. As we have explored throughout this guide, the practice of camel husbandry is not just about raising camels; it's a dynamic blend of tradition, science, and resource management that sustains livelihoods and cultures in some of the most challenging climates on Earth.

The camel's unique physiological traits, such as its ability to conserve water, withstand extreme temperatures, and traverse long distances with minimal sustenance, make it an ideal animal for desert and semi-desert regions. Effective camel husbandry involves understanding these traits and applying best practices in their care and management to maximize their productivity and welfare.

Successful camel husbandry requires a multifaceted approach. First and foremost, ensuring proper nutrition and health care is essential. Camels require a balanced diet tailored to their specific needs, which may include specialized feed, minerals, and access to clean water. Regular health check-ups and vaccinations are critical to prevent diseases that can devastate camel populations and impact their productivity.

Furthermore, breeding practices play a significant role in camel husbandry. Selecting camels with desirable traits, such as resilience to disease and adaptability to local conditions, helps improve the herd's overall performance and sustainability. Implementing effective breeding programs ensures the continuation of strong, healthy camel populations that can thrive in changing environments.

Cultural practices and traditional knowledge also play a significant role in camel husbandry. Many communities have developed sophisticated methods for managing camels, rooted in generations of experience. These practices should be respected and integrated with modern techniques to enhance productivity while preserving cultural heritage.

In addition, camel husbandry must address environmental and economic challenges. Sustainable practices that conserve resources and minimize environmental impact are crucial as climate change and land degradation affect pastoral lands. Economically, camel husbandry can offer a stable source of income through the sale of camel products like milk, meat, and wool, as well as tourism and transport services.

In conclusion, camel husbandry is a vital practice that supports both people and ecosystems in challenging environments. By combining traditional knowledge with modern advancements, camel husbandry can continue to provide economic stability, preserve cultural traditions, and sustain the unique role of camels in arid and semi-arid regions. As we move forward, embracing innovation while honoring tradition will ensure the continued success and sustainability of camel husbandry practices around the world.

THE END

www.ingramcontent.com/pod-product-compliance
Lightning Source LLC
Chambersburg PA
CBHW071835210526
45479CB00001B/144